1E. 69p

Oxford Graded
Arithmetic Practice

Problems 3

D. A. Holland
S. A. Soper

Oxford University Press
Walton Street Oxford OX2 6DP
OXFORD LONDON GLASGOW
NEW YORK TORONTO MELBOURNE
WELLINGTON KUALA LUMPUR
SINGAPORE HONG KONG TOKYO
DELHI BOMBAY CALCUTTA
MADRAS KARACHI NAIROBI
DAR ES SALAAM CAPE TOWN

ISBN 0 19 834733 2

Printed by The Blackmore Press,
Shaftesbury, Dorset

Oxford University Press 1980

Contents

Introduction

Word sums
Problems books 1, 2 and 3

Aims

1 To help the child to read, to understand, and to solve simple arithmetical problems.

2 To save time in the classroom by arranging material in graded examples for ease of use and continuity of method.

Structure

Word sums gives practice in working arithmetical questions set out in short phrases and in translating between words and symbols. It thus provides an introduction to the problems books, and it is also a useful aid to spelling.

The three problems books cover number, money, time, length, weight, capacity, simple fractions, decimals, and percentage. The careful grading takes account of difficulties in vocabulary, in comprehension, and in arithmetic. The wording of the questions aims for accuracy through the use of natural language that avoids ambiguity.

Use

These four books will be a useful supplement to any mathematics scheme since regular practice of reading and of the basic rules of arithmetic form an essential part of all courses at this level. Because the examples are graded and divided by pages into topics, it is a simple matter to isolate difficulties, to provide a controlled progression of work for each child, and to keep a record of individual progress.

Correction To simplify marking and analysis of errors, the layout of answer book and pupil's book is the same and where appropriate, children may correct their own answers.

Number

addition

1 50, 28, 14, 1479, 800 Add the largest of these numbers to the smallest.

2 In a school there are 697 boys and 514 girls. How many pupils are there altogether?

3 Two schools each sent half the pupils to a play. In one school there are 460 pupils and in the other 500 pupils. How many children saw the play?

4 Miss Hill made 15 egg sandwiches. She added one-third of these to the 8 ham sandwiches on the plate. How many sandwiches are now on the plate?

5 One-tenth of the 600 children in the school watched the football match. The opposing team brought 40 children to watch. How many children watched the match?

6 Team A has 30 points. Team B has one-third more points than Team A. How many points has Team B?

7 John has 86 stamps in his album. He bought 110 new stamps and stuck one-half of them in his album. How many stamps are now in the album?

8 Farmer Green sold one-third of his 90 cows and one-half of his 40 sheep. How many animals in all did he sell?

9 John had six sweets when Mary gave him one-quarter of her twelve sweets. How many sweets has John now?

Number

subtraction

1 In a school there are 297 boys and 342 girls. How many more girls are there than boys?

2 When two numbers are added together, the total is 870. If one of the numbers is 121, what is the other?

3 Jane made 69 sandwiches for a party. Two-thirds of them were eaten. How many sandwiches were left?

4 Ron lost one-quarter of his 164 marbles in a game. How many marbles had he left?

5 Sam took 144 paces to reach the end of the garden. James took 211. How many more paces did James take?

6 Team A has 143 points. Team B need 27 points to tie with them. How many points has Team B?

7 The sum of two decimal numbers is 116.8. If one of the numbers is 84.6, what is the other?

8 A cinema has seats for 1800 people. If 1495 people bought tickets for one performance, how many seats were left empty?

9 During an epidemic of 'flu, 374 of the school's 800 children were absent. How many children were present?

Number
multiplication

1 Five teams of eleven were picked from sixty boys. How many boys were left out?

2 Each page of Tom's album has space for 24 stamps. How many stamps will he have when he has filled 80 pages?

3 There are 78 children in the school who wear glasses. If this is one-quarter of the whole school, how many children are in the school?

4 At the station they collect about 465 tickets daily. About how many tickets will be collected from Monday to Friday?

5 Father gave out four hands of seven cards from a pack of 52. How many cards were left over?

6 A dress needs six buttons at the front and two for each sleeve. Nine of these dresses are to be made for the play. How many buttons are needed in all?

7 In a cricket match six balls make an over. How many balls are bowled in 13 overs?

8 Sue made a daisy chain using 40 daisies. She wants to make eight more daisy chains. How many more daisies does she need?

9 The machine drills 100 holes in an hour. How many holes can be drilled in a working day of $7\frac{1}{2}$ hours?

Number

division

1 What is the nearest number to 20 which will divide exactly by 3?

2 There were 40 children in Class 2. One in every four bought a ticket for the school concert. How many tickets were bought by Class 2?

3 At a meeting there are 72 adults. One-quarter of them are women. How many men are at the meeting?

4 Books come in packs of twenty. How many packs are needed to give 50 children two books each?

5 Two numbers are multiplied together and the answer is 78. If one number is 6, what is the other?

6 In a tray there are 30 eggs placed in rows of five. When half the eggs have been sold, how many rows can there be now?

7 Ten children were each given an equal share of a bag of 230 nuts. How many nuts did each child have?

8 There are 200 children in the school. Three-quarters of them went to the zoo in two equal parties. How many children were in each party?

9 Six children sit at each of the 5 tables in class. How many books are needed to give the children one book between two?

Number

mixed

1 Mary counted 96 lambs in the field with their mothers. How many mothers were there if each had 2 lambs?

2 Six posts are fixed in a row. How many spaces between posts are there?

3 What is the total if we add the smaller of 24 and 19 to the greater of 29 and 34?

4 If a batsman hits 2 sixes, 3 fours, and 15 singles, what is his score?

5 The houses in a row have odd numbers. The first is number one. What is the number of the ninth house?

6 Half the children in a group of 64 are given one rope between two. How many ropes are given out?

7 Three numbers added together make 21. If two of the numbers are 9 and 5, what is the other?

8 Find the difference between one-half of 22 and one-half of 38.

9 George and Bill each threw three darts. George scored nine, double six, and treble twenty. Bill scored double nineteen, seven, and double four. By how many did George win?

Money

addition

1 Mandy has £1.46 in her savings. Brian has £2.60 more than Mandy. How much money have they together?

2 In my purse I have 3 fifties, 4 tens, 6 fives, 2 twos, and 1 penny. What is the total value of these coins?

3 400p, £12.00, £150, 700p, £3.50 Add the largest amount to the smallest.

4 The collection amounted to £29.48, then a further £9.65 was added. What is the total now?

5 After paying a bill for £3.32 a man had 2 tens and 2 twos left. How much money had he at first?

6 Judy's weekly wage of £38.75 was increased by £1.70; what is her new weekly wage?

7 Add one-half of thirty-two pounds to one-half of thirty-eight pounds.

8 Last week my grocery bill was £15; this week it is 10% more. How much have I paid for groceries this week?

9 Mrs Gibson earns £78 per week. Her husband earns £12 a week more than she does. What is their combined weekly wage?

Money

subtraction

1 I received £1.12 change after paying for my meal with a five-pound note. How much had the meal cost?

2 At a sale, 10p in the pound was taken off all prices. What would be the actual cost of a mirror marked £8?

3 My aunt earns £75 each week. If she spends £18.75 of this on food, how much money is left from what she earns?

4 There is £10 in the cash tin. If Simon borrows £2.85 from it, how much is left in the tin?

5 Mrs Smith's fuel bill was £48.75 in the summer and £96.50 in the winter. What is the difference between these two amounts?

6 A black-and-white television set costs £128.40; how much less is this than a £400 colour set?

7 A dealer bought a vase for £275.50 and sold it for £500. How much profit did he make?

8 The single fare to town is £4.10 and the return fare is £7.50. How much cheaper is the return fare than two single fares?

9 How much is left out of £100 after spending £15.90 on a train ticket and £10 on a bag.

Money

multiplication

1 Father pays £1.40 in fares each day from Monday to Friday. How much will he pay for fares in two weeks?

2 Coffee is 23p per cup and buns are 12p each. Work out the cost of two coffees and three buns.

3 The bus fare to town was 20p. How much money did the driver collect from 15 passengers?

4 Material for curtains cost £8 per metre. Work out the price for $2\frac{1}{2}$ metres.

5 Each child in a class of 30 brought £1.05 for the school trip. How much money was collected altogether?

6 The cost of printing a notice was 40p per word. Calculate the cost of printing 30 words on the notice.

7 If Sally and Tom each save 25p weekly, how much do they save together in one year?

8 £100, 140p, £6.75, 300p, £20. Multiply the smallest of these amounts by five.

9 The ticket collector takes £2.24 from each of 6 children in a party. How much money does he collect from the whole party?

Money

division

1 How many coins should there be in a 50p bag of twopences?

2 The prize of £435 was shared between three people. How much money should each person receive?

3 How many 12p stamps can Harry buy for £7.32?

4 The charge for hiring a rowing boat is £1.50 per hour. If Jack was charged £6, for how long had he used the boat?

5 A man worked for eight hours and was paid £25.60 for the work. What was the hourly rate of pay?

6 Glasses cost 60p each. How many could be bought with eighteen one-pound notes?

7 Tim paid £4.80 for three model railway coaches. How much did each coach cost?

8 A herd of cows was sold for a total of £1252. How much should one-quarter of the herd have cost?

9 Mrs Moss paid £42.30 in fares for a group of five children to go to Birmingham. What was the fare for one child?

Money

mixed

1 If I buy 5 tenpenny stamps and 3 twelve penny stamps, how much change should I get from a one-pound note?

2 I paid £4.60 for three tins of fruit and a sack of potatoes. The fruit cost 60p per tin. How much did the potatoes cost?

3 Ten ball-point pens cost £1.10; what should have been the cost of three of the same pens?

4 In a sale all prices are reduced by one-third. How much should I now pay altogether for a £90 table and a £21 lamp?

5 Apples cost 60p per kilogram last week. This week the price has increased by 8p per kilogram. What is the cost of 3 kg of apples this week?

6 Debbie paid £2.30 for a sketch pad and 10 felt pens. The pad cost 80p. Work out the cost of one felt pen.

7 Harry spent one-third of his money and had 40p left. How much money did he spend?

8 How many $7\frac{1}{2}$p packets of gum could you buy with 45p?

9 Three cups of coffee and two 11p cakes cost 76p. Find the cost of two cups of coffee.

Time

addition

1 When his sister was born, John was 4 years old. Now his sister is two. How old is John?

2 If the 2nd of February is a Wednesday, what day of the week will be the 10th of February?

3 A lesson begins at 2.30 p.m. and lasts for 40 minutes. At what time does it end?

4 Rose went to bed at 10.00 p.m. She stayed there for 10 hours. At what time did she get up?

5 A 50-minute television programme started at 11.15 a.m. When did it end?

6 Twenty minutes have passed since the clock struck eleven. If the clock is 5 minutes fast, what is the time now?

7 Bob works from 9.00 a.m. till noon and from 1.15 p.m. till 5.00 p.m. How many hours is that each day?

8 Jill came home at 4.30 p.m. Before tea she did 40 minutes' homework and 20 minutes' piano practice. At what time did she have tea?

9 Mary was born 13 months before her brother who was born in January. In which month was Mary born?

Time

subtraction

1 Mr Wilson will retire in 11 years when he is 60. How old is he now?

2 Nina arrived at 4.40 p.m. after a journey of $5\frac{1}{2}$ hours. At what time had she set out?

3 A girl was 15 years old in 1979. In what year was she born?

4 A 35-minute lesson finished at a quarter to eleven. At what time did it begin?

5 My alarm clock is set to ring at a quarter past six. If I go to bed at a quarter past ten, how long will it be before the alarm rings?

6 The date on a Victorian penny is 1887. How old was this penny when it was found in 1976?

7 We left for our holidays on 19 July and came home on 26 July. For how many nights were we away?

8 Ron's time for the 800 m race was 6 min 42 s. Paul was faster by 55 seconds. What was Paul's time?

9 At 9 a.m. in London it is noon in Moscow. What is the time in London when it is 1.30 a.m. in Moscow?

Time
multiplication

1 A space satellite circles the Earth once every 90 minutes. How long does it take to circle the Earth eight times?

2 Tom spends 50 min every day tending his rabbits. Approximately how many hours is that in one week?

3 A lighthouse beacon flashes 12 times a minute. How often does it flash in one hour?

4 Mr Jones sleeps for one-third of every day. For how many hours will he sleep in the month of April?

5 Mother works 2 h 5 min each day in the kitchen. How much time does she spend in the kitchen in one week?

6 Each of 12 children spend 45 minutes collecting money for the disabled. How many hours is that altogether?

7 A cockerel crowed for about 10 minutes every morning. About how many hours did it crow in the month of May?

8 A clock gains 10 seconds every hour. How many seconds is that in one day?

9 If a wheel turns five times every second, how many turns will it make in one minute?

Time

division

1 A machine fills one milk bottle every 2 seconds. How many bottles will it fill in 2 minutes?

2 A pilot spent a total of 420 min on test flights. How many hours is that?

3 Tony's dog had to stay in quarantine for 168 days. How many weeks is that?

4 A clock gains 1 min 12 s every day. How many seconds does it gain in one hour?

5 A class of 25 children spent $1\frac{1}{4}$ hours on the trampoline. For how long could each child use the trampoline?

6 One man dug a trench in 2 hours. How long should it take four men to dig the same trench?

7 Dr Muller visited a patient once a fortnight for 2 years. How many times altogether did he visit her during this period?

8 John was on night duty one week in every four. For how many weeks in one year should he have been on night duty?

9 In a marathon contest a man stood on his head for 1200 seconds. How many minutes is that?

Time

mixed

1 Competitors leave at quarter-hour intervals. If the first man leaves at 1 p.m., at what time will the fifth man leave?

2 Jane was 7 years old on the 4th of July 1971. Write the date on which she became double this age.

3 A train is due in Liverpool at 11.55 p.m. If it is running 70 minutes late, at what time will it arrive in Liverpool?

4 Three years ago Mr Mills was 30. His daughter is now aged 12. How old was Mr Mills when his daughter was born?

5 School closed on 19 December and re-opened on 7 January. How long a holiday was this?

6 We left Oxford in the car at 10.00 a.m. We stopped only for lunch which took 40 minutes. If we arrived in Exeter at 3.00 p.m., for how long were we driving?

7 School starts at a quarter to nine but Jane arrived at three minutes past ten. How many minutes late was she?

8 Father's car was serviced three times a year. How many times was it serviced between the end of 1976 and the end of 1979?

9 Mr Adams works a 40-hour week. He has 9 weeks holiday in a year. About how many hours does he work in one year?

Length

addition

1 Ken painted $1\frac{1}{2}$ m of the plank red. Tom painted the remaining half blue. How long was the plank?

2 150 m, 25 cm, 2 km, 30 m, 250 cm Add the longest two distances together.

3 What is the distance all around a rectangular lawn that measures 40 m by 30 m?

4 The bookcase is 80 cm tall. It sits on the cupboard which is 1.40 m tall. What is the combined height?

5 A box is 8 cm wide and 3 cm deep. How much tape is needed to go round it once and make a 10 cm overlap?

6 The fence is 4.40 m long. The hedge is twice this length. What is the total length of hedge and fence?

7 Steven cycled 4.5 km. Paul cycled three times as far as Steven. How far did the two boys cycle altogether?

8 Joan beat the 3.95 m record for the long jump by 20 cm. What was the new long jump record?

9 We crossed the stream at 495 m above sea level and pitched the tent after climbing another 87 m. At what height above sea level was the tent?

Length

subtraction

1 Steven is 1.57 m tall. How far above his head is the ceiling which is 2.50 m from the floor?

2 A carpet is 4.50 m long. Its length is to be reduced by 0.34 m. What is the new length of the carpet?

3 Mary jumped 4.20 m in the long jump and beat Susie by 60 cm. How long was Susie's jump?

4 Mother cut off 120 cm from a piece of material 1.5 m long. How much material was left?

5 We began our journey in the car 170 km from Paris and stopped at a signpost which read Paris 75 km. How far had we travelled?

6 Paul's shadow was 2.9 m longer than his own height. How tall was Paul if the shadow was 4.1 m long?

7 A climber is one-third of the way up a 1500 m mountain. How much higher must he climb to reach the summit?

8 The school corridor is 48 m long. If one-quarter of it is shut off for redecoration, what length if left open?

9 Mary travelled 2470 km on her holiday. I travelled 590 km less than Mary. How far did I travel?

Length
multiplication

1 Ten sheets of 6 mm thick plywood and ten sheets of 12 mm thick plywood are stacked in a pile. How many centimetres thick is the pile?

2 One side of a square field measures 42.5 m. Calculate the length of the perimeter of this field.

3 Simon walked a total of 600 m to and from school each day. How many kilometres is that in one week?

4 Eight planks each 3.5 cm thick are placed one on top of the other. What is the height of the pile?

5 An athlete runs round the 500 m track seven times during a training session. How many kilometres does he run altogether?

6 If a plant grows 10 mm each week, how many centimetres will it grow in one year?

7 Peter's tree is eight times his own height. If Peter is 1.4 m tall, find the height of the tree to the nearest metre.

8 Jim's father has a set of twelve drain rods, each 90 cm long. How far will the rods reach if they are all joined together?

9 A man uses 2.4 m of wire to make a frame. How many metres of wire does he need to make twelve similar frames?

Length

division

1 Thirty similar planks are laid end to end across a muddy field. If their total length is 150 m, how long is each plank?

2 The perimeter of a square field is 152 m. What is the length of each side?

3 How many 40 cm lengths of string can be cut from a 12 m ball of string?

4 The distance all around a running track is 500 m. How many times will an athlete run around the track in a 2 km race?

5 I am to knit a scarf 1.80 m long. If I knit 15 cm each day, how long will I take to knit the scarf?

6 Fred walked a total distance of 132 km on holiday. Mary walked one-quarter of this distance. How far did Mary walk?

7 Mrs Hunt cut up an old washing line into 9 equal lengths to make skipping ropes. If the line was 22.5 m long, how long was each rope?

8 Twelve equally spaced dots make a line 11 cm long. What is the distance between one dot and the next?

9 One-quarter of a 2.4 m pole is cut off. Write in centimetres the length of pole that is left.

Length

mixed

1 The sides of a triangle measure 3 cm, 3.5 cm and 2.5 cm. Find one-half of the perimeter of this triangle.

2 A fence 9 m long is supported by ten equally spaced posts. How far is the third post from the first?

3 A rectangular tile measures 14 cm by 8 cm. How much longer than 40 cm is the perimeter of this tile?

4 When a piece of paper is folded in half, it forms a 10 cm square. What is the size of the paper?

5 A dictionary is 20 mm thick. How many of these dictionaries would form a pile 30 cm high?

6 A doorway is 75 cm wide. How much space does that give each side of a trolley 39 cm wide?

7 The school record for the high jump was 1.10 m. Tom jumped 6 cm higher than this. How much higher still would he have to jump to set a new record of 1.20 m?

8 Four 25 cm pieces are cut off a length of wood. This leaves 1 metre. How long was the wood at first?

9 The width of a table cloth is one-third of its length. If the cloth is 1.8 m long, how wide is it?

Weight

addition

1 130 g, 22 g, 5 kg, 31 g, 2 kg Add the heaviest of these weights to the third heaviest.

2 I bought 1 kg of apples, 500 g of tomatoes, 250 g of dried fruit, and 250 g of tea. What is the total weight in kg?

3 A tin of peas weighs 540 g and a tin of carrots weighs 280 g. Calculate, to the nearest kilogram, the total weight of 2 tins of peas and 3 tins of carrots.

4 In a box of fruit gums there are 36 g of sweets. The box weighs 4 g. What percentage of the total weight is the box?

5 Jack's suitcase weighs 18 kg. He adds a pair of shoes weighing 1.2 kg and a shirt weighing 500 g. What is the weight of his suitcase now?

6 On Tuesday the school had collected 1800 kg of old newspapers. A further 500 kg was collected on each of the next three days. What was the total weight in tonnes?

7 Tim weighs 24 kg and his sister weighs twice as much as Tim. How much do they weigh together?

8 Mother bought 500 g of apples, 500 g of plums and 2 kg of sugar. What percentage of the total weight was fruit?

9 600 g butter, $1\frac{1}{2}$ kg flour, 400 g sugar, 50 g salt, 50 g ginger. Find the total weight to the nearest kilogram.

Weight

subtraction

1 Father weighs 75 kg and Mother weighs 59.7 kg. Find the difference in their weights.

2 A sack contains $16\frac{1}{2}$ kg of corn. If one-third of the corn is sold, what weight of corn remains in the sack?

3 A joint of meat weighs 800 g. If the bone weighs 200 g, what percentage of the joint is meat?

4 What is the difference in weight between a 5 kg sack of potatoes and two 750 g bags of potatoes?

5 Molly weighed 70 kg. After slimming she now weighs 63 kg. What percentage of her weight has she lost?

6 A man carrying a 17.7 kg suitcase weighs 89.2 kg. How heavy is the man?

7 A lorry carried 750 kg of coal. When two-thirds of this had been delivered, what weight of coal was left on the lorry?

8 Judy's luggage weighed 31.8 kg but she cannot take more than 25 kg on the aeroplane. What weight of luggage must she leave behind?

9 A lorry load of sand weighed $1\frac{1}{2}$ tonnes. One tonne of this sand was delivered. What percentage of the load remained on the lorry?

Weight

multiplication

1 Tom weighs 23.4 kg. His father is three times as heavy as Tom. What is the weight of Tom's father?

2 A baker ordered four sacks of flour. If each sack weighed 27.5 kg, find the total weight of his order.

3 A box of matches weighs 10.7 g. Find the weight, to the nearest gram, of a pack of 12 of these boxes.

4 Ten boxes each contain 3.7 kg of tomatoes. Find the total weight of tomatoes.

5 Mother usually buys 4 kg of fruit and 6 kg of vegetables every week. About what weight of fruit and vegetables will she buy in one year?

6 James eats 80 g of bread daily. How many kilograms of bread will he eat during the month of September?

7 Thirty children each picked $4\frac{1}{2}$ kg of raspberries. What weight is that altogether?

8 How many one-kilogram weights would be needed to balance five bags of cement, each weighing 1.4 kg?

9 What weight of fruit gums is there in a pack of 20 tubes if there are 9 fruit gums in each tube and one fruit gum weighs 4 grams?

Weight

division

1 How many 500 g weights would be needed to balance a 20 kg box of steel rods?

2 A bucket contains 15 kg of wet sand. If $33\frac{1}{3}\%$ of the weight is water, find the weight of sand.

3 A large fruit cake weighed 2.85 kg. If it was cut into 5 equal pieces, how much would each piece weigh?

4 Seven sacks of carrots weighed 43.4 kg. How much would one sack weigh?

5 Food for hens is delivered in 25 kg bags. If a farmer ordered one tonne of hen food, how many bags should he receive?

6 Mother made a total of 12 kg of jam. She put it into 8 jam pots. What weight of jam did each pot hold?

7 A sack of sugar contains 16.8 kg. If the sugar is shared equally between 12 people, how much does each person receive?

8 Mr Forbes the chemist divided 4 g of aspirin equally between 8 bottles. Find the weight, in milligrams, of aspirin in each bottle.

9 A scoop can hold $\frac{1}{4}$ kg of flour. How many scoopfuls are needed to put 2.25 kg of flour into a bag?

Weight

mixed

1 What is the difference in weight between ten 12 g packets and one $\frac{1}{2}$ kg packet?

2 A lorry carried 20 sacks of coal each weighing 50 kg. When three-quarters of the coal had been delivered, what weight was left?

3 Richard picked $2\frac{1}{2}$ kg of fruit, Tom picked 500 g of fruit and Murray picked 2 kg of fruit. How many 500 g baskets did they fill?

4 One parcel weighs twice as much as another. Together they weigh 9 kg. Find the weight of each parcel.

5 How many bags of potatoes, each weighing 3 kg, can be made up from a delivery of 75 kg?

6 Three boxes of plums together weigh 11 kg. If the empty boxes each weigh 250 g, what is the weight of plums?

7 A lorry is carrying 620 kg less than its maximum load of 2 tonnes. What weight is the lorry carrying?

8 The total weight of 6 tins of fruit is 6.360 kg. Find the total weight of four of these tins.

9 Miss White took 11 ball pens from the box. Altogether they weighed 66 g. She gave two of the pens to John who put them on the scales. What weight should the scales show?

Capacity

addition

1 Add one-quarter of a litre to the total of these quantities: 350 ml, 570 ml, 1 litre, 830 ml.

2 John made 4.5 litres of lemonade. Mark made twice as much. How much did they make altogether?

3 How many litres of fruit cocktail are made from 400 ml of lemon, 600 ml of orange, 600 ml of pineapple and 800 ml of water?

4 Ann used 13 litres of water but Sue used only half as much. How much water did they use altogether?

5 Bill has two 10 litre buckets containing water. One is half-full and the other is three-quarters full. How many litres of water is that?

6 The stove used 6 litres of oil in November and half as much again in December. How much oil was used in the two months?

7 Mary bought 2 litres of green paint, $1\frac{1}{4}$ litres of blue paint, 6 litres of white paint and 500 ml of black paint. How much paint had she bought in all?

8 The tank collected 2400 litres of water in February and 1800 litres in March. How much is that altogether?

9 At school there are three fish tanks. They hold 4.5 litres, 6 litres, and 8.4 litres of water. How much water, to the nearest litre, would be needed to fill all three tanks?

Capacity

subtraction

1 How much more paraffin should be added to a 10 litre tank to fill it if it already contains 4.2 litres?

2 Tim used 9.8 litres of paint mixed with water. If he used 4.5 litres of water, how much paint did he use?

3 Of the 36 kilolitres of milk delivered to a dairy, one-quarter was returned. How many kilolitres of the milk were not returned?

4 The tank contained 600 litres of water. Mr Finch drained off 200 litres from the tank. What percentage of the water remained?

5 A petrol tanker carried 12 000 litres of petrol. What quantity of petrol had been delivered when the tanker returned with its tank one-third full?

6 A swimming pool containing 2000 litres of water was to be drained. After one hour, three-quarters of the water had gone. How many litres were left?

7 What is the difference in capacity between a 500 litre container and a 5000 litre container?

8 A chemist took 100 ml of medicine from a one-litre jar. What percentage of the medicine was left in the jar?

9 A flask contained 3 litres of squash, but 10 per cent of this had leaked through the cap. How much squash was left?

Capacity
multiplication

1 If a lamp uses 40 ml of oil in a day, how many litres of oil will it burn during April?

2 Mary fills thirty 200 ml bottles with water. How many litres of water does she need?

3 Twenty $1\frac{1}{2}$ litre jugfuls of water are used to fill a tank. If Alan completely filled the tank twice, how much water did he use?

4 Mr and Mrs Young each drink 600 ml of milk every day. How much milk, to the nearest litre, do they drink each week?

5 Mother used 10 bowls of water, each containing 7.4 litres, to fill the paddling pool. How much water did she use in all?

6 A machine fills twenty 100 ml bottles with oil in one minute. How many litres of oil are bottled in one hour?

7 I bought 10 cans of drink for our journey. Each can contained 160 ml. How many litres of drink did I buy altogether?

8 Ten petrol tankers each carried 20 000 litres of petrol. How many kilolitres is that?

9 A tap leaked 1.5 ml in one minute. How much water escaped in one hour?

Capacity

division

1 Linda takes a daily dose of medicine for 10 days. In all she takes 150 ml. How much is each daily dose?

2 Mrs Green divides 2 litres of squash equally between 8 cups. How much will she pour into each cup?

3 A mountaineer had used one-quarter of his 3.6 litre ration of water. How many millilitres had he used?

4 Father shared one-half of the $2\frac{1}{2}$ litres of milk between his five children. How much did each child have?

5 Martin was asked to share 5 litres of juice equally between 25 cups. How much juice should he put in each cup?

6 500 ml, 3.6 litres, 36 ml, 930 ml. Divide the greatest of these quantities by three.

7 How many 150 ml glasses could be filled from a jug containing 1.5 litres?

8 During a water shortage a tanker brought 2000 litres of water for the 20 houses in our road. How much water is that for each house?

9 Peter had 1 litre of squash. He kept 30 per cent of it and shared the remainder between 4 friends. How much squash did each friend have?

Capacity

mixed

1 How many 5 litre jars could be filled from twenty 2 litre bottles?

2 Each of 180 children drank one-quarter of a litre of lime juice from a 50 litre barrel. How much was left in the barrel?

3 Father took one-third of the 12 litres of fertilizer and mixed it with an equal quantity of water. How much mixture did he make?

4 When Paul had used up one-third of his 3.9 litre ration of water, how much water had he left?

5 Mother mixed $1\frac{1}{2}$ litres of water with the same amount of orange juice and 500 ml of lemon. She shared this mixture between 7 children. What amount did each child receive?

6 Father's car used 150 litres of petrol in April, and half as much in May. How much petrol did the car use in the two months?

7 John made 12 litres of ginger beer. He filled four 2 litre bottles with the ginger beer and wasted the rest. What percentage of the ginger beer was wasted?

8 Tom's cat drank 400 ml of milk every day. Tom drank twice as much. How much milk was needed altogether in one week?

9 Sheila made 3 litres of milk shake. Half of this she put in the fridge. The rest she shared equally with her four friends. How much milk shake did each of the five girls have?